AF540443

Advanced Techniques in Plant Genetic Resources

About the Authors

Dr. S. Prabakaran has done his B.Sc. (Agri.) from TNAU sub-campus at Trichy. He studied his M.Sc. and Ph.D. from Indian Agricultural Research Institute (IARI), New Delhi. His area of specialisation was Plant Genetic Resources. In 2015, he received ICAR-JRF 22nd rank in plant sciences and secured merit fellowship. In 2017, he secured 1st rank IARI entrance exam and secured IARI merit fellowship for Ph.D. program. He also qualified Agricultural Scientist Recruitment Board (ASRB) – National Eligibility Test (NET) in 2017. He has qualified UGC-SRF and lectureship in Environmental sciences in 2019. Currently, he is teaching agriculture optional to civil service aspirants at Chennai.

A. Krishnamoorthi I did my Bsc agriculture(2012-16) from VIA Agriculutre college Affiliated to TNAU and I did my Msc Agriculture in Plant geneticr esource (2108-20) from TNAU and I am written 10 book chapters 2 research article and 10 review articles I cleared my ICAR SRF Examination with Rank 4 in the Discipline of Economic Botany and Plant Genetic Resources. Now I am pursuing my Ph.D. at the ICAR-National Bureau of Plant Genetic Resourrces IARI Pusa Campus New Delhi.

Advanced Techniques in Plant Genetic Resources

S. Prabakaran
A. Krishnamoorthi

1168, Sector 13, Urban Estate
Karnal-132 001, Haryana
Tel: 91-84470 75807, 18440 41168
Email: contentvibesppa@gmail.com
www.contentvibes.in

Print ISBN: 978-93-6134-834-1

ebook-ISBN: 978-93-6134-599-9

Preface

The growing global population and the challenges posed by climate change underscore the importance of sustainable agricultural practices. The foundation of these practices lies in the genetic diversity of plant resources, which is essential for breeding resilient and productive crops. "Advanced Techniques in Plant Genetic Resources" addresses this critical need by providing an in-depth exploration of the latest methodologies and technologies used in the conservation, characterization, and utilization of plant genetic resources.

This book is structured to serve as a comprehensive guide for researchers, students, and professionals in the fields of plant science, genetics, and agriculture. It brings together contributions from leading experts worldwide, presenting a blend of theoretical foundations and practical applications. Our goal is to offer readers both the knowledge and tools needed to navigate and contribute to the dynamic landscape of plant genetic resource management.

The chapters cover a broad spectrum of topics, including:

- **Conservation Strategies:** *In situ* and ex situ conservation methods, cryopreservation, and seed banks.
- **Characterization Techniques:** Molecular markers, genomic sequencing, and phenotyping technologies.
- **Utilization Approaches:** Breeding strategies, biotechnological interventions, and the development of genetically modified organisms (GMOs).
- **Data Management:** Bioinformatics tools, databases, and information systems for managing genetic resource data.
- **Policy and Ethical Considerations:** Intellectual property rights, access and benefit-sharing agreements, and the socio-economic impacts of genetic resource utilization.

Throughout this book, we emphasize the importance of an integrated approach that combines traditional knowledge with cutting-edge scientific advancements. We also highlight the significance of international collaboration and policy frameworks in ensuring the sustainable use and conservation of plant genetic resources.

We are grateful to the contributors who have shared their expertise and insights, making this book a valuable resource for the global scientific community. We also extend our appreciation to the institutions and organizations that support research and development in plant genetic resources.

As you delve into the chapters that follow, we hope you will find inspiration and practical guidance that will enhance your work and contribute to the broader goal of sustainable agriculture and food security. Together, through the application of advanced techniques and a commitment to conservation, we can harness the full potential of plant genetic resources to meet the challenges of the 21st century.

Authors

Contents

1

Principles of Cryopreservation Techniques

Abstract

Preservation of genetic resources is very important because, until two decades ago genetic resources were getting depleted owing to the continuous depredation by the man. It was imperative therefore that many of the elite economically important and endangered species were preserved to make them available when you need. The conventional methods of storage failed to prevent losses caused due to various reasons. A new methodology has to be devised for long term preservation of material.

Keywords: *Cryopreservation, Liquid State, Solid State*

Cryopreservation

Cryo is a Greek word (krayos-frost). It literally means preservation in frozen state. **Cryopreservation** is the non lethal storage of biological material at ultra-low temperatures. At the temperature of liquid nitrogen (-196°C) almost all metabolic activities of cells are ceased and the sample can then be preserved in such state for extended periods. This is to bring plant cells or tissues to a zero metabolism and non-dividing state by reducing the temperature in the presence of cryoprotectant. However only few biological materials can be frozen without affecting the cell viability.

It can be done by using,

- Over solid carbon dioxide (at -79 degree)
- Low temperature deep freezer(-80 degree)
- In vapour phase nitrogen(at -150 degree)
- In liquid nitrogen(at -196 degree)

Among these, liquid nitrogen is most widely used material for cryopreservation because it is chemically inert, relatively low cost, Nontoxic, Non-flammable and Readily available.

Steps involved in cryopreservation

1. Selection of plant material
2. Pregrowth
3. Addition of cryopreservants
4. Vitrification
5. Cryopreservative dehydration
6. Encapsulation and dehydration
7. Freezing
8. Rapid freezing
9. Slow freezing
10. Step wise freezing
11. Storage
12. Thawing
13. Determination of survival or viability

Mechanism of cryopreservation

It is based on the transfer of water present in the cells from liquid to solid state. Due to presence of cells in the organic molecules in the cell, the cells water require much more lower temperatures to freeze compared to freezing point of pure water. when stored at low temperature, the metabolic process, biological deteriorations in the cells/tissues almost comes to standstill.

Selection of plant material

Morphological and **physiological** conditions of plant material influence the ability of explants to survive during cryopreservation. Different types of tissues can be used for cryopreservation such as :

- Ovules
- Anther/pollen
- Embryos
- Endosperm
- Protoplast, etc.

Factors responsible for selection of plant material

- Tissue must be selected from healthy plants
- It can be young, small and rich in cytoplasm.
- Meristematic cells can survive better than the larger cells and highly vacuolated cells

- A rapidly growing stage of callus shortly after 1 or 2 weeks of subculture is best for cryopreservation.
- Callus derived from tropical plant is more resistant to freezing damage.
- Old cells at the top of callus and blackened area should be avoided.
- Cultured cells are not ideal for freezing, Instead, organized structures such as shoot apices, embryos, young plantlets are preferred.
- Water content of cell or tissue should be low freezable water, tissues can withstand extremely low temperatures.

Pre-growth

- Pre-growth treatment protect the plant tissues against exposure to liquid nitrogen
- Involves the application of additives known to enhance plant stress tolerance.

  Eg: Abscissic acid, Praline, trehalose
- Partial tissue dehydration can be achieved by the application of osmotically active compounds.
- The addition of low concentration of DMSO (1-5%) during pre growth often improves shoot tip recovery.

  Eg:
- *C. roseus* cells are pre cultured in medium containing 1M sorbitol before freezing (Chen *et al.*, 1984)
- *Digitalis* cells were precultured on 6% mannitol medium for 3 days before freezing (Seitz *et al.*, 1983)
- *Nicotiana sylvestris* with 6% sorbitol for 2-5 days before freezing (Maddox *et al.*, 1983)

Addition of a cryopreservant

- A cryopreservant is a substance that is used to protect biological tissue from freezing damage(due to ice formation).
- They act like antifreeze
- They lower freezing temperatures
- Increase viscosity and
- Prevents damage to the cells
- There are two potential sources of cell damage during cryopreservation
- Formation of larger ice crystals inside the cell.

- Intracellular concentration of solutes increase to toxic levels before or during freezing as a result of dehydration,

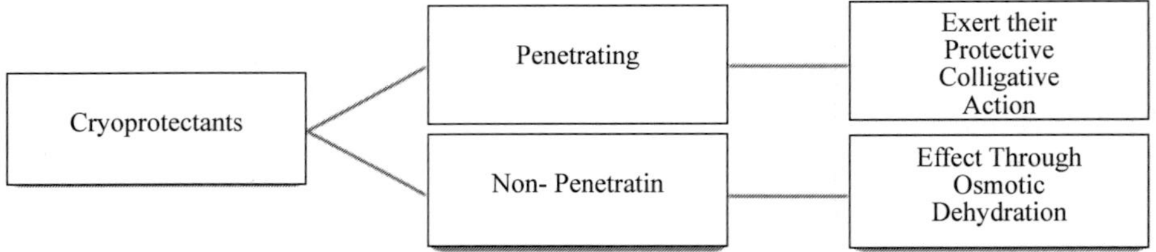

Vitrification

Vitrification is the plate transition of water from liquid to a non crystalline or amorphous glass state in cryopreservation. The formation of a metastable glass during exposure to liquid nitrogen effectively avoids injury to th tissue caused by intracellular ice crystallisation. Plant vitrification solution (PVS2) was developed by Sakai. Vitrification involves three major phases like loading phase, vitrification phase and unloadiong phase.

a) **Loading phase:** Pretreating the tissue in a cryoprotectant mixture for a short period to cryoprotect the tissues and to enable them to withstand osmotic stress and chemical toxicity.

b) **Vitrification:** Tissue is dessicated in a vitrification mixture for a short period before vitrifiying in liquid nitrogen.

c) **Unloading phase:** Rapid warming there by vitrification solution drained out of the cryogenic vials and replaced with sucrose at elevated concentration.

There are eight steps ionvolved in vitrification preocedure

Procedure

Step 1: Selection of explants

Step 2: Preculture in cryoprotectants

Step 3: Pretreatment with loading solution (2M glycerol+0.4M sucrose or 10% glycerol)

Step 4: Dehydration by PVS2 (plant vitrification solution 2 =30% glycerol +15% ethylene glycol+15% DMSO + 0.4M sucrose)

Step 5: Cryopreservation and storage in liquid nitrogen

Step 6: Thawing

Step 7: Removing PVS2 (unloading with high sucrose concentration)

Step 8: Regeneration of plants.

Encapsulation – vitrification

This method combines the encapsulation and vitrification technique establisged by Engelmann and was reported first by Matsumoto et al., using shoot apices. In this method samples are encapsulated in alginate beads then subjected to freezing by vitrification.

Materials suitable: shoot apices, callus, pollen

Procedure

1. Selection of explants
2. Encapsulation in sodium alginate with cryopreservation
3. Dehydration in PVS2
4. Cryopreservation and storage in liquid nitrogen
5. Thawing
6. Regeneration

Droplet vitrification

This method was first reported by Schafer-Menuhr et al., using potato.

1. Selection of explants
2. Pre culture in vitrification solution
3. Cryopreservation in droplets of PVS2 placed in aluminium foil strips
4. Storage in liquid nitrogen
5. Regrowth and regeneration

Dehydration

This was first reported by Urgani *et al.*, 1990 using asparagus lateral buds.It is a simple procedure since consists of dehydrating explants,then freezing them rapidly by direct immersion in liquid nitrogen.

Materials suitable - Zygotic embryos, embryogenic axes extracted from seeds.

Procedure

1. Selection of explants
2. Dehydration using silica gel or air flow for 60-360 min (250 degree)
3. Cryopreservation and storage
4. Thawing
5. Regeneration

Encapsulation-dehydration

This technique is based on development of synthetuc seed developed by Fabre and Deruddre. This involves the **encapsulation** of tissues in **calcium alginate beads.**

Materials used- Somatic embryos, shoot tips, bilblets, nodal sements, callus, cell suspension

- Which are pre-grown in liquid culture media containing high concentration of sucrose.
- After these treatments the tissues are able to withstand exposure to liquid nitrogen without application of chemical cryoprotectants.

Freezing

Freezing is done in such a away that it does not cause any intracellular freezing and crystal formation.

The following are the types of freezing used in the process of cryopreservation:

- **Slow freezing method:** initially the temperature frozen is at the rate of 0.5°C-4°C per minute from the temperature of 0°C and allowed to reach -100°C and then transferred to liquid nitrogen. This is followed by peas, strawberry, potato, cassava etc. This method is highly useful for preserving cells from suspension cultures.
- **Rapid freezing method:** The material is frozen rapidly by plunging into liquid nitrogen thereby the temperature decreases at the rate of -300 to -1000°C per minute. Dry-ice (CO_2) can also be used instead of liquid nitrogen. Rapid freezing has been employed to cryopreservation of shoot tip carnation, strawberry, potato and *Brassica napus*. This method is recommended to specimens with low water content.
- **Droplet freezing:** In this method the cryoprotectant treated meristem are dispensed in droplets of 2-3 micro litre of aluminium foil in a petriplate. The specimens are frozen by slow cooling (0.5°C minute) to sub-zero temperature between -20°C to 40°C prior to immersion in liquid nitrogen.
- **Stepwise method:** The plant material is subjected to slow freezing rate initially (1 to 5°C per minute) for 30 minutes and afterwards rapidly cooled by plunging into liquid nitrogen.
- **Dry freezing method:** In this method the plant material are dehydrated by drying in an oven or under vacuum to reduce cryogenic damage.

Storage

- Storage of frozen material at correct temperature is as important as freezing. The frozen cells/tissues are kept for storage at temperature ranging from -70 to -196°c. Temperature should be sufficiently low for long term storage of cells to stop all the metabolic activities and prevent biochemical injury.
- Long term storage is best done at -196°c.

Thawing

- It is done by putting ampoule containing the sample in a warm water bath (35 to 40°c).
- Frozen tips of the sample in tubes or ampoules are plunged into the warm water with a vigorous swirling action just to the point of ice disappearance.
- It is important for the survival of the tissue that the tubes should not be left in the warm water bath after ice melts .
- just a point of thawing quickly transfer the tubes to a water bath maintained at room temperature and continue the swirling action for 15 sec to cool the warm walls of the tube.
- Tissue which has been frozen by encapsulation/dehydration is frequently thawed at ambient temperature.

Determination of survival /viability

- Regrowth of the plants from stored tissues or cells is the only test of survival of plant materials.
- Various viability tests include Fluorescien diacetate (FDA) staining, growth measurement by cell number, dry and fresh weight.

Important staining methods are

- Triphenyl Tetrazolium Chloride (TTC)
- Evan's blue staining.

Measurement of growth of cell cultures

- Fresh and dry weight measurements
- Increase in cell number
- Packed cell volume (PCV)
- Molecular protien and DNA

- Mitotic index
- Medium component calibration
- Conductivity of medium
- Cellular protien

Fresh and dry weight measurement

- For callus
- Transfer the entire callus to pre-weighed weighing boat.
- Determine the fresh weight.

For cell suspension culture

- Collect the cells on a pre weighed nylon membrane
- Determine the fresh weight

Increase in cell number

- Haemocytometer can be used.
- Callus or cells suspension is needed to be macerated
- Maceration fluid is an equal volume of 10% chromic acid and 10% nitric acid.
- Treat this mix for 5 to 10 minutes at 60 degree.
- After cooling, shake the container vigorously to loosen the cell clumps.
- 1% macerozyme incubate flask overnight under growth condition.

Packed cell volume

- Suitable for suspension culture only.
- Simple and can be determined at any stage of growth.
- A small volume (10ml) is aseptically sampled and placed in graduated conical centrifuged tube.
- Total volume of a cell pellet is determined after centrifuging at a constant speed of (500g) for specific time (5 mins).
- PCV is typically expressed as a percentage of the total volume in tube.

Molecular protein and DNA

- The components like DNA and protien are usually determined by using well established techniques

  e.g. Barford's reagent for protien.

Mitotic index

- Mitotic index is an estimate of the number of cells of population in the stages like prophase, metaphase, anaphase and telophase.
- It is simple but time consuming.

  (Number of nuclei in mitosis/ total number of nuclei scored) × 100

Freeze storage of cell cultures

- A cell line to be maintained has to be subcultured and transferred periodically and repeatedly over an extended period of time.
- cryopreservation is an ideal approach to suppress cell division to avoid the need for periodical subculturing.

Applications

- Preservation of rare genomes
- Freeze storage of cell cultures
- Conservation of genetic uniformity
- Maintenance of disease free material which is ideal for the international exchange
- Cold acclimation and frost resistance
- Retention of morphogenetic potential
- Slow metabolism which would prevent or virtually 'stop' the ageing process.

Preoblems in cryopreservation

1. Solution effects

These are caused by concentration of solutes in non frozen solution during freezing as soluted excluded from the crystal structure of ice. High concentration of solution is very dangerous.

2. Extracellular ice formation

When tissues are cooled slowly, water moves out of cells and ice formation occurs in extra cellular space. Too much extra cellular ice formation causes damage due to crushing.

3. Dehydration

Migration of water causing extra cellular ice formation also causes cellular dehydration. This results in stress which causes damage directly to cell.

4. Low viability or no regrowth after cryopreservation

This is one of the drawback, which is due to problems in preculturing step, in the cryopreservation step, at freezing and thawing steps.

5. Cryovial rupturing

Liquid nitrogen is trapped in vials, evaporated rapidly and increased pressure in the vials results into rupturing of cryovials.

6. Browning

Due to phenolics, inappropriate subculturing time resulting in the death of many cells.

Advantages and disadvantages of different cryopreservation methods

Cryopreservation method	Advantages	Disadvantages
Vitrification	• Requires less handling • Eliminates the need for controlled slow freezing • Does not need expensive cooling apparatus • Avoid ice formation during cooling • More flexibility	Toxicity caused by cryoprotectants Direct exposure of tissues to PVS 2 reduces viability
Encapsulation-Vitrification	• Doesn't need plant vitrification solution • Low cost	Ready influence of humidity on drying air flow
Dehydration	• Easy and low cost • Useful for many temperate tree crops	Require expensive freezing equipment Larger storage space
Encapsulation- Dehydration	• Not expensive • Toxicity caused by cryopretectants especially DMSO is eliminated. • Doesn't need expensive cooling apparatus	Larger storage space
Pregrowth	• Simple method • No special equipment required	Only applicable for winter hardy buds, seeds or zygotic embryos
Pregrowth- dehydration	Simple and easy method	Special equipment required Larger storage space

Conclusion

- Adopted in pathogen-eradication schemes with high efficiency for species and genotypes that are going to be cryopreserved.
- This technique based procedures require some special equipment.
- The use of vitrification protocols and optimized regeneration of shoots will minimize the risk of somaclonal variation.
- Important method in conservation of crops at extinct state.

References

Engelmann, F. (2004). Plant cryopreservation: progress and prospects. In vitro cellular & developmental biology-Plant, 40, 427-433

Pegg, D. E. (2002). The history and principles of cryopreservation. In Seminars in reproductive medicine (Vol. 20, No. 01, pp. 005-014). Copyright© 2002 by Thieme Medical Publishers, Inc., 333 Seventh Avenue, New York, NY 10001, USA. Tel.:+ 1 (212) 584-4662.

Pegg, D. E. (2009). -Principles of Cryopreservation. PreservAtion of HumAn oocytes, 34-46.

2

Legislations and Acts of PGR Access and Benefit Sharing

Abstract

The current international debate on legal regimes for plant genetic resources has its origins in the late 1970s and early 1980s when industrialized countries started granting patents on improved plant varieties and the developing countries became concerned over such extension of intellectual property rights to the varieties

Introduction

This led to an extended debate and international co-operation in the recognition of plant-related intellectual property rights, resulting in a greater attention being paid to questions of ownership of genetic resources. At the same time concerns for sustainable use of biological diversity were also raised.

One of the major events was the Convention on Biological Diversity (CBD) which came into force in 1993, which was adopted during the Rio Earth Summit of the United Nations. It was the first legally binding institutional mechanism, providing for conservation and sustainable use of all biological diversity and intends to establish the process of the equitable sharing of benefits arising out of the use of biodiversity.

The Convention on Biological Diversity

The CBD was conceived in the 1980s and was negotiated under the leadership of the United Nations Environment Programme (UNEP). It was adapted in June 1992 at Rio de Janeiro, at the Earth Summit, and is effective since December 1993.

The objectives of CBD are

- The conservation of biological diversity,
- The sustainable use of its components and
- The fair and equitable sharing of benefits arising out of utilization of genetic, including by appropriate access to genetic resources and by appropriate transfer of relevant technologies, taking into account all

rights over those resources and to technologies, and by appropriate funding.

The convention reaffirmed that states have sovereign rights over their biological resources and that the states are responsible for conserving these resources and using them in a sustainable manner. The countries that signed the CBD were required to integrate considerations of sustainable use of biological diversity into relevant sectoral or cross-sectoral plans programmes and policies.

Access and Benefit Sharing

1. Monetary benefits

a) Access fees/fee per sample collected or otherwise acquired
b) Up-front payments
c) Milestone payments
d) Payment of royalties
e) License fees in case of commercialization
f) Special fees to be paid to trust funds supporting conservation and sustainable use of biodiversity
g) Salaries and preferential terms where mutually agreed
h) Research funding
i) Joint ventures
j) Joint ownership of relevant IPRs.

2. Non-monetary benefits

a) Sharing of research and development results
b) Collaboration, cooperation and contribution in scientific research and development programmes, particularly biotechnological research activities, where possible in the provider country
c) Participation in product development
d) Collaboration, cooperation and contribution in education and training
e) Admittance to *ex situ* facilities of genetic resources and to databases
f) Transfer to the provider of the genetic resources of knowledge and technology under fair and most favourable terms, including on concessional and preferential terms where agreed; in particular, knowledge and technology that make use of genetic resources, including biotechnology, or that are relevant to the conservation and sustainable use of biological diversity

g) Strengthening capacities for technology transfer to user developing country Parties and to Parties that are countries with economies in transition and technology development in the country of origin that provides genetic resources. Also to facilitate abilities of indigenous and local communities to conserve and sustainably use their genetic resources
h) Institutional capacity building
i) Human and material resources to strengthen the capacities for the administration and enforcement of access regulations
j) Training related to genetic resources with the full participation of providing Parties and, where possible, in such Parties
k) Access to scientific information relevant to conservation and sustainable use of biological diversity, including biological inventories and taxonomic studies
l) Contributions to the local economy
m) Research directed towards priority needs, such as health and food security, taking into account domestic uses of genetic resources in provider countries
n) Institutional and professional relationships that can arise from an access and benefit-sharing agreement and subsequent collaborative activities
o) Food and livelihood security benefits
p) Social recognition
q) Joint ownership of relevant IPRs.

The Biological Diversity Act, 2002

The Biological Diversity Act of India (BDA) was formulated after India became signatory to the CBD in 1993. It was developed through an intensive consultation process involving central government, state governments, institutions of local self-government, scientific and technical institutions, experts, non-governmental organisations, industry etc. This act was passed by the Parliament in December 2002, and has the following objectives based on the principles of CBD.

i. To regulate access to biological resources of the country with the purpose of securing equitable share in benefits arising out of the use of biological resources, and associated knowledge relating to biological resources
ii. To conserve and sustainably use biological diversity
iii. To respect and protect knowledge of local communities related to biodiversity

iv. To secure sharing of benefits with local people as conserves of biological resources and holders of knowledge and information relating to the use of biological resources

v. Conservation and development of areas important from the standpoint of biological diversity by declaring them as biological resources

vi. Protection and rehabilitation of threatened species

vii. Involvement of institutions of self-government in the broad scheme of the implementation of the Act through constitution of committees.

Management Structure of Biodiversity Act

A three tiered structure at the national, state and local level is envisaged.

National Biodiversity Authority (NBA)

All matters relating to requests for access by foreign individuals, institutions or companies, and all matters relating to transfer of results of research to any foreigner will be dealt with by the National Biodiversity Authority.

State Biodiversity Boards (SBB)

All matters relating to access by Indians for commercial purposes will be under the purview of the State Biodiversity Boards (SBB). The Indian industry will be required to provide prior intimation to the concerned SBB about the use of biological resource. The State Board will have the power to restrict any such activity, which violates the objectives of conservation, sustainable use and equitable sharing of benefits.

Biodiversity Management Committess (BMCS)

Institutions of local state government level of Panchayats, Zilla Taluks, Municipalities, will be required to set up biodiversity management Committees in their respective for the purpose of

- Promoting Conservation,
- Sustainable use
- Documentation of biological diversity including
  - Preservation of habitats
  - Conservation of land races, folk varieties, cultivars domesticated stocks and breeds of animal and microorganisms
  - Chronicling of knowledge associated to biological diversity.

Access And Benefit Sharing

The Act stipulates norms for access to biological resources and traditional knowledge in three ways:

i. Access to biological resources and traditional knowledge to foreign citizens, companies and non-resident Indians (NRIs) based on 'prior approval of NBA'.
ii. Access permits to Indian citizens, companies, associations and other organizations registered in India on the basis of 'prior intimation to the State Biodiversity Board' concerned.
iii. Exemption of prior approval or intimation for local people and communities, including growers and cultivators of biodiversity, and Vaids and Haqims, practicing indigenous medicines.

These benefits sharing includes

a) Grant of joint ownership of intellectual property rights to the NBA, or where benefit claimers are identified, to such benefit claimers
b) Transfer of technology
c) Location of production, research and development units in such areas which will facilitate better living standards to the benefit claimers
d) Association of Indian scientists, benefit claimers and the local people with research and development in biological resources and bio-survey and bio-utilization
e) Setting up of venture capital fund for aiding the cause of benefit claimers
f) Payment of monetary compensation and other non-monetary benefits to the benefit claimers as the NBA may deem fit.

TRIPS and its benefit sharing

Plant protection was also affected by the Uruguay Round of the General Agreement on Trade and Tariffs (GATT) of 1986. The GATT is a multilateral agreement intended to reduce international trade barriers. An important component of the GATT is settlement of trade-related aspects of intellectual property rights (TRIPs). The GATT has moved closer to more universal recognition of plant breeders' rights. The GATT creates minimum standards for the protection of intellectual property rights over commercially developed seed and plant varieties. Article 27, 3(b) states that, "Members shall provide for the protection of plant varieties either by patents or by an effective *sui generis* system or by any combination thereof."

UPOV (Union for Protection of Plant Varieties) and benefit sharing

UPOV aims to maximize plant breeding efforts by providing a model for securing protection under UPOV for plant breeders' rights for plant varieties. It was enacted in 1968 but amended in 1972, 1978 and 1991. The UPOV Convention incorporates the principle of free access to improved varieties for further research and breeding (breeders' exemption). In its present form, the UPOV model would exclude the imposition of a requirement to disclose the origin of genetic resources as a condition for the granting of PBR.

ITPGRFA (International Treaty on Plant Genetic Resources for Food and Agriculture)

This Treaty will enter into force on June 29, 2004. Currently, there are 144 contracting parties. It includes over 35 food crops and 29 forages species i.e, 64 crops which contributes 85 per cent of global human nutrition. The Treaty applies only to plant genetic resources useful for food and agriculture (PGRFA). It establishes the following objectives:

1. To encourage the conservation of plant genetic resources in order to preserve and enhance the genetic diversity of plant species and varieties of value to food or agriculture
2. To provide a workable, juridical basis for rewarding farmers for their contributions in conserving, improving, and making available plant genetic resources
3. Further development of the system of national sovereignty over plant genetic resources first established in the CBD, while ensuring that such exercise of sovereignty does not hinder international exchange of such resources
4. Creation of a Multilateral System of Access and Benefit-Sharing, which will coordinate exchanges of plant genetic resources, and in some cases, require payments by persons or entities who commercially exploit such resources, to the nations from which such resources originated.

Access and Benefit Sharing

1. **The exchange of information**: this includes catalogues and inventories, information on technologies and results of technical, scientific and socio-economic research on PGRFA including data on characterization, evaluation and information on use.
2. **Access to and transfer of technology**: Contracting Parties agree to provide or facilitate access to technologies for the conservation, characterization, evaluation and use of PGRFA. The ITPGRFA lists various means by which transfer of technology is to be carried out, including participation in crop-based or thematic networks and partnerships, commercial joint ventures, human resource development and through making research facilities available. Access to technology, including that protected by IPR, is to be provided and/or facilitated under fair and most-favourable terms, including on concessional and

preferential terms where mutually agreed. Access to these technologies is provided while respecting applicable property rights and access laws.

3. **Capacity building**: the ITPGRFA gives priority to programmes for scientific education and training in the conservation and use of PGRFA, to the development of facilities for conserving and using PGRFA and to the carrying out of joint scientific research.
4. **Sharing of monetary and other benefits arising from commercialization**: monetary benefits include payment into a special Benefit-Sharing Fund of the MLS of a share of the revenues arising from the sale of PGRFA products that incorporate material accessed from the MLS.

PPV & FRA (Protection of Plant Varieties and Farmers' Rights Act)

It is an effective system for the protection of plant varieties, the rights of farmers and plant breeders, to encourage the development of new varieties of plants.

The **objectives** of the Act are as follows:

i. To provide for the establishment of an effective system for protection of plant varieties.
ii. To provide for the rights of farmers and plant breeders.
iii. To stimulate investment for research and development and to facilitate growth of the seed industry.
iv. To ensure availability of high quality seeds and planting materials of improved varieties to farmers.

Period of Protection

i. In the case of trees and vines, eighteen years from the date of registration of the variety;
ii. In the case of extant varieties, fifteen years from the date of the notification of that variety by the Central Government under Section 5 of the Seed Act, 1996, and
iii. In the other case, fifteen years from the date of registration of the variety.

Benefit sharing

Sharing of benefits accruing to a breeder from a variety developed from indigenously derived plant genetic resources has also been provided. The authority may invite claims of benefit sharing of any variety registered under the Act, and shall determine the quantum of such award after ascertaining the extent and nature of the benefit claim, to both the plant breeder and the claimer.

Nagoya Protocol

It was enacted in 29 October 2010 in Nagoya, Japan but entered into force during 12 October 2014. It aims at sharing the benefits arising from the utilization of genetic resources in a fair and equitable way

- By appropriate access to genetic resources
- By appropriate transfer of relevant technologies,
- By appropriate funding, thereby contributing to the conservation of biological diversity and the sustainable use of its components

It also covers traditional knowledge (TK) associated with genetic resources . It also addresses genetic resources where indigenous and local communities have the established right to grant access to them. It provides incentives for the promotion and protection of traditional knowledge.

Access and Benefit-sharing

- Create conditions to promote and encourage research contributing to biodiversity conservation and sustainable use
- Pay due regard to cases of present or imminent emergencies that threaten human, animal or plant health
- Consider the importance of genetic resources for food and agriculture and their special role for food security
- Establish clear rules and procedures for prior informed consent and mutually agreed terms
- Provide fair and non-arbitrary rules and procedures
- Provide for the fair and equitable sharing of benefits arising from the utilization of genetic resources, as well as subsequent applications and commercialization, with the contracting party providing the genetic resources
- Ensure that sharing of benefits is subject to mutually agreed terms.
- Benefits may be monetary (such as royalties) or non-monetary (such as sharing research results or technology transfer)

References

Brahmi, P., & Tyagi, V. (2015). Policy Issues in PGR Management. Management of Plant Genetic Resources, National Bureau of Plant Genetic Resources, New Delhi, 323 p. Published in 2015 All Rights Reserved ICAR-National Bureau of Plant Genetic Resources New Delhi 110012, India Copies can be obtained from: The Director ICAR-National Bureau of Plant Genetic Resources New Delhi 110012, India E-mail: director@ nbpgr. ernet, 248.

Engels, J. M. M., Withers, L., Raymond, R., & Fassil, H. (2001). The importance of PGR and strong national programmes. Towards Sustainable National Plant Genetic Resources Programmes: Policy, Planning and Coordination Issues, 1, 2001.

Ghose, J. R. (2003). Access and Benefit Sharing Systems: An Overview of the Issues and the Regulation.

3

Germplasm Introduction and Exchange

Abstract

Germplasm is vital resource in generating new plant types having desired traits that help in increasing crop production and thus improve the level of human nutrition. It fuels all the activities of modern agriculture - research, improvement and production. The plant breeders need more diverse germplasm so that the acquisition of superior varieties by importing them from other areas has now become absolutely necessary.

Keywords: *Germplasm exchange, Primary introduction, Secondary introduction, import permit*

Introduction

According to Frankel (1957), plant introduction is the transposition of a genetic entity from an environment to which it is attuned to one in which it is untried. According to Bennett (1965), the introduction of wild plants into cultivation and the successful transfer of cultivars, with their genotypes unaltered, to new environments is called as 'primary' plant introduction and the rest as 'secondary' introduction. However, both are equally meaningful for effective use of plant germplasm.

Types of Plant Introduction

1. Primary Introduction
2. Secondary Introduction

1. Primary Introduction

When the introduced crop or variety is well suited to the new environment, it is directly grown or cultivated without any alteration in the original genotype. This is known as primary introduction. E.g. IR. 8, IR 20, IR 34, IR 50 rice varieties; oil palm varieties introduced from Malaysia and Mashuri rice from Malaysia.

2. Secondary Introduction

The introduced variety may be subjected to selection to isolate a superior variety or it may be used in hybridization programme to transfer some useful traits. This is known as secondary Introduction. E.g. In soybean EC 39821 introduced from Taiwan is subjected to selection and variety Co 1 was developed. In rice ASD 4 is crossed with IR 20 to get Co 44 which is suited for late planting.

Objectives of Plant Introduction

- To introduce new plant species there by creating ways to build up new industries. E.g. Oil
- Palm
- To introduce high yielding varieties to increase food production. E.g. Rice and wheat.
- To enrich the germplasm collection. E.g. Sorghum, Groundnut.
- To get new sources of resistance against both biotic and abiotic stresses. Dasal rice variety for saline resistance.
- Aesthetic value – ornamentals are introduced for aesthetic value.

Plant Introduction Agencies

Most of the introductions occurred very early in the history. In earlier days the agencies were invaders travellers, traders, explorers, pilgrims and naturalists. In India, Muslim invaders introduced cherries and grapes. Portuguese introduced maize, ground nut, chillies, potato, sweet potato, guava, pine apple, papaya and cashew nut. East India Company brought tea. Later Botanic gardens played a major role in plant Introduction.

Plant Introduction Agencies in India	
NBPGR - National Bureau of Plant Genetic Resources (New Delhi)	Agriculture & Horticultural crops
FRI - Forest Research Institute (Dehradun)	Forestry plants
BSI - Botanical Survey of India (Calcutta)	Plants of Botanical Interest

NBPGR- A centralized plant introduction agency and as a single window system is responsible for introduction and maintenance of germplasm of agricultural and horticultural plants. History of NBPGR

- Initiated in 1946 at IARI in the Division of Botany, ICAR, New Delhi.
- In 1956, under second five year plan, expanded as Plant Introduction and Exploration organization
- In 1961, upgraded as independent Division of Plant Introduction, IARI, New Delhi.
- In 1976, upgraded to an independent institute as National Bureau of Plant Introduction (NBPI) In 1977, renamed as NBPGR

Functions of NBPGR

1. Introduction maintenance and distribution of germplasm
2. Provide information about the germplasm through regular publications.
3. Conduct training courses to the scientists for introduction and maintenance of germplasms.
4. Conduct exploratory surveys for the collection of germplasm.
5. To set up Natural gene sanctuaries.At International level International Board of Plant Genetic Resources (IBPGR) with head quarters at Rome, Italy is responsible for plant introduction between countries.

Procedure for Plant Introduction

The scientist / University will submit the requirement to NBPGR. If the introduction is to

be from other countries, NBPGR will address IBPGR for effecting supply. The IBPGR will assign collect the material from the source and quarantine them, pack them issue phytosanitary certificate suitably based on the material and send it to NBPGR. The NBPGR will assign number for the material, keep part of the seed for germplasm and send the rest to the scientist. There are certain restrictions in plant introduction. Nendran banana from Tamil Nadu should be not be sent out of state because of bunchy top disease. Similarly, we cannot import Cocoa from Africa, Ceylon, West Indies, Sugarcane from Australia, Sunflower from Argentina.

Steps in Plant Introduction

1. Procurement
2. Quarantine
3. Cataloguing
4. Evaluation
5. Multiplication

1. Procurement of germplasm

The new germplasm is procured through NBPGR, New Delhi. Scientists, individuals and institutions can submit their requirement to Director, NBPGR, Pusa Complex N.Delhi-12. If the bureau is unable to meet the request from its own stock or from known source it attempts to procure them from the counterparts in other countries. Generally the material is obtained through correspondence as gifts or exchange of germplasm in consideration of past gifts to the Bureau or in anticipation of future gifts. The plant part depending on the crop species e.g. seeds; tubers; suckers, bulbs or cuttings etc. can be procured.

2. Quarantine

It is to keep the material in isolation to prevent spreading of diseases etc. All introduced material is thoroughly inspected for contamination with diseases, weeds and insects. Plants that are suspected to be contaminated are fumigated or are given other isolation for treatments and observed for insect pests and disease. It is essential that all the material being introduced must be accompanied by an authentic phytosanitary certificate. The plant material being introduced or exported must confirm to certain quarantine regulations and quarantine control is exercised by NBPGR at different points of entry. The phytosanitary certificate is thoroughly inspected and returned back to the sender or owner.

3. Cataloguing

All the plant material which is introduced is given an entry number and information regarding agency, place of origin, adaptation etc and is well documented. Plant material is classified in three categories

- Exotic collection (EC)
- Indigenous collection (IC)
- Indigenous wild collection (IW)

4. Evaluation

The plant material is sent to sub stations of the bureau and evaluated with respect to various characters to assess the potential of new introductions.

5. Multiplication and distribution

Plant material which is introduced is to be multiplied and further tested at various locations. The suitability of cultivation in different regions of the country should be assessed before using it as a commercial variety.

Acclimatization

When superior cultivars from neighbouring or distant regions are introduced in a new area, they

generally, fail initially to produce a phenotypic expression similar to that in their place of origin. But later on they pickup and give optimal phenotypic performance, in other words they become acclimatized to the new ecological sphere. Thus, acclimatization is the ability of crop variety to become adapted to new climatic and edaphic conditions.

The process of acclimatization follows an increase in the frequency of those genotypes that are better adapted to the new environment. The success of acclimatization depends upon two factors

i. Place effect
ii. Selection of new genotypes.

Merits of plant introduction.

1. It provides new crop varieties, which are high yielding and can be used directly
2. It provides new plant species.
3. Provides parent materials for genetic improvement of economic crops.
4. Enriching the existing germplasm and increasing the variability.
5. Introduction may protect certain plant species in to newer area will save them from diseases.

   E.g. Coffee and Rubber.

Demerits

1. Introduction of new weed unknowingly. E.g. *Argemone mexicana*, *Eichornia* and *Parthenium*
2. Introduction of new diseases:Late blight of potato from Europe, Bunchy top of banana from SriLanka
3. New pests: Potato tuber moth came from Italy, woolly aphid of apple, fluted scale of Citrus
4. Ornamentals becoming weeds: *Lantana camara*
5. Introduction may cause ecological imbalance E.g. *Eucalyptus*.

Procedure for Exchange of Germplasm

NBPGR, New Delhi has brought out a brochure 'Guidelines for the exchange of seed/planting material' (National Bureau of Plant Genetic Resources, 1986) and this had been widely circulated among scientists in India.

Import

All requests for indenting germplasm from abroad are to be made to the NBPGR giving specific details of the required material, stating the source/country as well as address of organisation/scientist through a prescribed application so that the Import Permit is issued and sent to concerned scientist(s) for sending the same to exporting organisation or the Bureau be requested to persue the import of desired material(s) from abroad. When requests are sent directly by an individual scientist to any foreign source without an 'Import Permit', the

NBPGR needs to be kept informed of such requests for the issuance of 'Import Permit'. Two mandatory requirements for import of germplasm

- Import permit issued by NBPGR before import of any material (IP)
- Phytosanitary certificate from the country of origin (PSC)

Issuance of Import Permit

- Director, NBPGR - authorized to issue import permit and receive imported materials from customs authorities for its quarantine inspection and clearance, as per clause 6 (2) of PQ Order 2003.
- The recipient desirous of importing seed/planting material has to apply to the Director, NBPGR on a prescribed application form (PQ Form 08).
- The IP is issued in form PQ 09 in triplicate.
- IP is valid for six months from the date of issue and valid for successive shipment provided the exporter and importer, bill of entry, country of origin and phytosanitary certificate are the same for the entire consignment.
- Validity may be extended up to one year on request, if adequate reasons in writing are justified.
- Import permit is non-transferable
- IP to be sent to the supplier and he should attach the same in duplicate along with the consignment

Issuance of Import Permit for Transgenics

- An indentor who wishes to import transgenics has to submit the proposal to Review Committee on Genetic Manipulation (RCGM) through the Institutional Biosafety Committee (IBSC).
- RCGM is an authorized agency of the GOI, functioning under DBT, which assesses the applications submitted for importing transgenic material for research purposes and issues Seed Transfer Clearance Letter.
- RCGM examines the desirability of import of transgenic line, from the biosafety point of view.
- It includes representatives from DBT, ICMR, ICAR, CSIR and other experts in their individual capacity.
- After getting the technical clearance for import, from RCGM, the indentor has to apply to Director, NBPGR, New Delhi for the issuance of IP in the prescribed PQ 08 form along with Seed Transfer Clearance Letter issued by DBT

Release of the Consignment

- After obtaining the IP, the recipient should send it to the concerned official/supplier (abroad) who has agreed to supply the required germplasm for use in research.
- The supplier should be instructed to send the copies of IP (one pasted outside the seed parcel and other inside the seed packet) and other relevant documents with the consignment.
- The IP and PSC in duplicate must be enclosed along with the seed/ planting material for custom clearance.
- It should be clearly instructed to the sender/supplier that consignment should be addressed only to the Director, NBPGR, New Delhi.
- The port of entry of germplasm is New Delhi Airport only.

Registration, National Accessioning and Import Quarantine

- All indents for import of germplasm are registered for assigning the case number and then forwarded to the Plant Quarantine (PQ) Division, without opening the parcel along with duly filled Import Quarantine (IQ) form, for detailed quarantine inspection and clearance.
- After clearance from PQ Division, the samples are first arranged taxonomically showing genus, species, common name and cultivar name etc. for national accessioning in the national record.
- Each introduction/accession is assigned an Exotic Collection (EC) number which remains unchanged with information like name and address of the indentor, address of the source institute/country, characteristics and details of the germplasm, relevant references, date of arrival, form of the material and recipients of the materials.
- All assembled healthy seed/plant material is dispatched to the various researchers to make use of these valuable genetic resources
- The accessioning of the imported material is done online at www.nbpgr.ernet.in/geq.
- Only small qty imported. For transgenics – DBT specifies qty.; 5 – 10 % retained as voucher sample

The concerned scientist/organisation abroad is advised to take into consideration the following requirements for mailing the material to India.

1. Only healthy, viable and clean seed material (free from soil, pests, pathogens and weeds) are to be forwarded without any seed treatment so as to facilitate proper quarantine examination. It may, however, be fumigated, if considered necessary.

2. The material is required to be accompanied with an 'Import Permit' and 'Phytosanitary Certificate' with additional declaration, if any, based on crop inspection certifying that the material is free from particular pathogen(s)/insect(s). Preferably, two copies of the phytosanitary certificate, one inside the package and the other affixed outside for use by the quarantine authorities are required. It should further be ensured that the package of seed/planting material must be addressed to the Director, NBPGR, New Delhi, who has to take delivery of the seed/ planting material and conduct quarantine examination.
3. The seed/planting material should not be sent to any scientist by name directly, since one point entry is necessary to have proper monitoring for quarantine requirements, for documentation of passport data and national accessioning (E.C. Number assignment referring to Exotic Collection).
4. The seed may preferably be sent by first class air mail addressed to the Director, National Bureau of Plant Genetic Resources, Pusa Campus, New Delhi - 110012.
5. The perishable plant propagules (scion/woods, budwoods, plant rhizomes, suckers, etc.) may preferably be sent by airfreight through any commercial airlines operating between source country and the Indira Gandhi International Airport, New Delhi so as to avoid delay in receipt and clearance. If unavoidable, the material can be sent on charge collect basis. An advance intimation of the despatch of such perishable materials to the NBPGR will help in prompt receipt and quick clearance of the material soon after its arrival. This will also avoid payment of demurrage charges. However, in case it is not possible to send by air freight, the same could be sent by courier service.
6. The sender should be requested to give full particulars of the seed/ planting material along with the address of the concerned scientist to whom the material is to be made available after quarantine clearance by the Bureau.
7. Germplasm should be obtained in small quantities (3000 to 4000 seeds only) and, in case of vegetative materials, it should not exceed six scion woods, rhizomes, etc., while in case of rooted plants, the number should be kept to the minimum (1-2 plants each). Efforts should be made to avoid repeat introductions. When requests are routed through the NBPGR, this aspect is taken care of since supply of the required seed/ planting material may possibly be arranged from sources within India, depending on its availability.

Export

Exchange of germplasm involves not only introductions but also the supply of seed and other materials to collaborating scientists/organisations abroad. While responding to such requests, the following guidelines are to be followed:

1. Requests for seed/planting material received from concerned organisations/agencies abroad have to be forwarded to the NBPGR with relevant information so that prompt decision on the supply of desired materials could be taken.
2. Despatch of the seed/planting materials is also to be channelized through the Bureau so that prompt inspection of the material could be done from quarantine angle and phytosanitary certificate be issued.
3. Only the healthy seed material (free from diseases, pests, weeds, soil clods, plant debris, etc.) should be sent to the Bureau in small quantity along with full details of the material(s) and the name and address of the recipient. Quarantine clearance and despatch normally takes about 7 to 10 days.
4. No seed dressing with insecticides or fungicides be given while despatching the seed to the Bureau.

Export of Plant Genetic Resources

- The access to Indian biological resources by foreign nationals or export of germplasm for research purposes is governed by Biological Diversity Act (BDA), 2002 & Biological Diversity Rules (BDR), 2004 (http://www.biodiv.org). The act was enacted under the provisions of the CBD.
- As per section 3 of the Act, no person from outside India or a corporate, association, organization incorporated or registered in India having non -Indian participation in its share capital or management, can access any biological resources or associated knowledge, for research , commercial utilization, bio-prospecting or bio-utilization, without prior approval of National Biodiversity Authority (NBA).
- However, as per Section 5 of the Act, exchange of germplasm for collaborative research under the bilateral agreements/collaborative projects are, exempted, which confirm to the policy guidelines issued by the central govt. or approved by the central govt.
- The proposals for export received under collaborative research projects are examined and screened by the Germplasm Export Facilitation Committee (GEFC).

Documents to be submitted along with export proposal

- Request Letter from foreign institution submitted to ICAR or through Research Institutions or any researcher in India or outside India
- Details of the Collaborative Research Project duly signed.
- Details of the seed/planting material under consideration for export
- Import permit of the recipient country
- Signed copy of ICAR/DARE approved Material Transfer Agreement (MTA)/ Standard Material Transfer Agreement (SMTA)
- Proforma for Export (check list) duly filled in and signed

- The GEFC submits its recommendations to DARE, for the final approval.
- After the approval is conveyed by Subject Matter Division (SMD), the seed/planting material is collected from NBPGR/NAGS and PSC is issued by PQ Division after quarantine inspection of the seed/planting material meant for export.
- The seed/planting material is dispatched to the indentor along with IP and PSC

Indian Varieties Developed Using Exotic Material Introduced by ICAR-NBPGR (2012 – 2018)

Crop	Variety [Exotic material used]
Field Pea (Pisum sativum var. arvense)	Harbhajan [EC-33866]
Soybean (Glycine max)	Monetta [EC-2587]
Sunflower (Helianthus annuus)	K-1 [Introduction from Russian cultivar EC 68414]
Groundnut (Arachis hypogaea)	Type-64 (Virginia group) [Sel. From EC 16064]
Barnyard millet (Echinochloa colonum L.)	PRJ-1 [Selection from IEC 542 (Collection of ICRISAT)]

Conclusion

Taking a genotype or a group of genotypes in to a new place or environment where they were not grown previously. Thus, introduction may involve new varieties of a crop already grown in that area, a wild relative of the crop species or totally a new crop species for that area.It may involve new varieties of crop already grown in that area, wild relatives of the crop species or a totally new crop species for the area. Often the term is used for introducing the material from other countries, but movement of crop varieties from one environment into another within the country is also termed as introduction.

References

Allard, R.W. 1960. Principles of plant breeding. John Wiley & Sons. London.

Bennett, E. 1965. Plant introduction and genetic conservation: Genecological aspect of an urgent world problem. Scottish Plant Breed. Sta. Res., pp. 27-113.

Chang, T.T. 1987. Saving crop germplasm. Span (Feb. issue) 62-63.

Dadlani, S.A., B.P. Singh and R.V. Singh. 1981. System of national and international exchange of germplasm and methods of recording followed at NBPGR. Sci. Monogr. No. 5, NBPGR, New Delhi. pp. 72-87.

Frankel, O.H. 1957. The biological system of plant introduction. J. Aust. Inst. Agric. Sci. 20: 302-7.

4

Role of CGIAR in Germplasm Conservation and Exchange

Abstract

Stands for '**Consultative Group for International Agricultural Research**'. *It is a global partnership that unites international organizations engaged in research for a food-secured future. CGIAR research is dedicated to reducing rural poverty, increasing food security, improving human health and nutrition, and ensuring sustainable management of natural resources.*

Keywords: *rural poverty, increasing food security, improving human health and nutrition.*

Introduction

It is carried out by 15 centres that are members of the CGIAR Consortium, in close collaboration with hundreds of partners, including national and regional research institutes, civil society organizations, academia, development organizations, and the private sector.

History

It was formed in response to the widespread concern in the mid-20th century that rapid increase in human populations would soon lead to widespread famine. The Rockefeller Foundation and the Mexican government, in 1943 laid the seeds for the Green Revolution when they established the Office of Special Studies, which led to establishment of the International Rice Research Institute (IRRI) in 1960 and International Maize and Wheat Improvement Centre (CIMMYT) in 1963. However, these foundations alone could not fund all the agricultural research and development efforts needed to feed the world's population. In 1969, the Pearson Commission on International Development urged the international community to undertake "intensive international effort" to support "research specializing in food supplies and tropical agriculture". Thus, with the support of World Bank, FAO and UNDP, CGIAR was established on May 19, 1971, to coordinate international agricultural research efforts aimed at reducing poverty and achieving food security in developing countries

CGIAR initially supported

- CIMMYT
- IRRI
- The International Centre for Tropical Agriculture (CIAT)
- The International Institute of Tropical Agriculture (IITA)

Their initial focus was on rice, wheat and maize. By 1970s, it included cassava, chickpea, sorghum, potato, millets and other food crops, and encompassed livestock, farming systems, the conservation of genetic resources, plant nutrition, water management, policy research, and services to national agricultural research centres in developing countries. By 1983, there were 13 research centres around the world under its umbrella. By 1990s, a total of 15 institutes came under the guidance of CGIAR.

Vision of Cgiar

"To reduce poverty and hunger, improve human health and nutrition, and enhance ecosystem resilience through high-quality international agricultural research, partnership and leadership".

Objectives of Cgiar

CGIAR's vision is supported by 4 strategic objectives:

- Reducing rural poverty
- Improving food security
- Improving nutrition and health
- Sustainably managing natural resources

Organisations Under Cgiar

1. AfricaRice Centre

- Since its inception in 1970, AfricaRice has been involved in genetic resources activities, specifically collection, preservation, characterization, evaluation, multiplication and distribution of promising germplasm to rice scientists in Africa and throughout the world.

  Germplasm collection in West Africa by both international and national research centres was intensified in the 1980s.
- Germplasm collected includes:
  - *Oryza sativa*
  - *O. glaberrima*
  - *O. longistaminata*
  - *O. barthii*
  - *O. stapfii*.

- From 1985 to 1993, AfricaRice have collected over 6000 accessions consisting of *O. sativa* (4800), *O. glaberrima* (1200), *O. longistaminata* (10), *O. barthii* (6) and *O. stapfii* (3). New collections of landraces were made in Côte d'Ivoire and Guinea.
- To strengthen genetic-resources activities, Africa Rice formally created a Genetic Resources Unit in 1999, within which the International Network for Genetic Evaluation of Rice for Africa (INGER-Africa) is nested
- Today, the gene bank contains close to 20,000 rice accessions. Long-term conservation is undertaken at IITA (Ibadan, Nigeria) and medium-term at AfricaRice's temporary headquarters in Cotonou, Benin.
- Duplicates are planned to be kept in Fort Collins, USA.
- Now the online portal, Africa Rice Genebank Information System (ARGIS) is available on Internet.
- ARGIS contains information on germplasm conserved in AfricaRice's Genebank, how to obtain seeds from Africa Rice using Standard Material Transfer Agreement (SMTA) and many other datasets. The database holds records of over 19,000 rice accessions.

2. Bioversity International

- Their vision – that agricultural biodiversity nourishes people and sustains the planet
- It's headquarters are in Rome, Italy, with regional offices located in Central and South America, West and Central Africa, East and Southern Africa, Central and South Asia, and South-east Asia
- Bioversity International maintains the world's largest banana gene bank, the "Bioversity International *Musa* Germplasm Transit Centre", that is hosted at the *Katholieke Universiteit Leuven* in Leuven, Belgium, and manages **ProMusa** - a platform that shares knowledge about bananas and plantains
- In 2002, the Global Crop Diversity Trust was established by Bioversity International on behalf of the CGIAR and FAO with the help of Crop Diversity Endowment Fund
- Since it's establishment, the Crop Trust has funded work in over 80 countries, and made its first grant for long-term conservation of rice in late 2006.
- By 2011, the Crop Trust had established in-perpetuity support (i.e., grants funded through the Crop Trust's endowment fund) for collections

of 15 crops which included rice, cassava, wheat, barley, faba bean, pearl millet, maize, forages, banana, aroids, grass pea, sorghum, yam and lentil

- The Crop Trust joined the Government of Norway and the Nordic Gene Bank in the 2008 for the establishment of the Svalbard Global Seed Vault, a "fail-safe" facility located at Svalbard, Norway.

3. Centre for International Forestry Research (CIFOR)

- The Centre for International Forestry Research (CIFOR) is a non-profit scientific research organization that conducts research on the use and management of forests with a focus on tropical forests in developing countries
- It's headquarters are located in Bogor, Indonesia
- CIFOR is a global knowledge organization committed to enhancing the benefits of forests and forestry for people
- The Centre was established in response to global concerns about the social, environmental and economic consequences of loss and degradation of forests.
- The six thematic work areas are:
    - Forests and human well-being
    - Sustainable landscapes and food
    - Equal opportunities, gender, justice and tenure
    - Climate change, energy and low-carbon development
    - Value chains, finance and investments
    - Forest management and restoration
- The CGIAR Research Program on Forests, Trees and Agroforestry (FTA) is the world's largest research for development program to enhance the role of forests, trees and agroforestry in sustainable development and food security and to address climate change

4. International Centre for Agricultural Research in Dry Areas (ICARDA)

- The **International Centre for Agriculture Research in the Dry Areas (ICARDA)** is a non-profit agricultural research institute that aims to improve the livelihoods of the resource-poor across the world's dry areas.
- ICARDA has been temporarily headquartered in Beirut, Lebanon, since leaving Aleppo, Syria in 2012

- Established in 1977, ICARDA's origin lies in a 1973 study that highlighted the food security challenges faced by countries across the dry regions of the Near East and North Africa
- ICARDA's gene bank was established in 1985 in Tel Hadiya, Syria
- It contains 1,57,000 samples of major winter cereals, food legumes, forage and rangeland species drawn from four major Vavilovian centers of plant diversity – including the 'Fertile Crescent' in Western Asia, the Abyssinian highlands in Ethiopia, and the Nile Valley, where the earliest known crop domestication practices were first recorded
- Many plants are now extinct or endangered in their natural habitats.
- ICARDA's collection of unique genetic materials rank among the most important worldwide. These collections are rich in landraces and wild relative species. More than 85 percent of the accessions are characterized and 75 percent are geo-referenced. All accessions are safely duplicated in reliable gene banks around the world and around 75 percent are duplicated at the Svalbard Global Seed Vault
- ICARDA also receives accessions from other gene banks and donors around the world. It receives genetic stocks and elite breeding germplasm with known valuable traits from breeders and researchers. All of these acquisitions are done using the FAO Germplasm Acquisition Agreement or Standard Material Transfer Agreement
- It covers **durum and bread wheat, barley, chickpea, lentil, faba bean, grasspea, and forage and pasture crops**
- One of the key tasks of ICARDA's gene bank is to respond to requests for accessions to ensure continuum between the conservation and utilization of genetic resources

5. International Centre for Tropical Agriculture (CIAT)

- Non-profit research and development organization dedicated to reducing poverty and hunger while protecting natural resources in developing countries
- Headquarters are located in Palmira, Colombia
- The main conservation and breeding strategies are carried out for crops like beans, cassava and tropical forages

1. Beans

- CIAT safeguards the world's largest and most diverse collection of bean germplasm

- The collection consists of around 36,000 samples of cultivated materials mostly from the crop's Mesoamerican and Andean centres of origin together with wild species related to these materials. The CIAT collection constitutes a valuable resource for bean improvement worldwide.

2. Cassava

- CIAT conserves in vitro the world's most important collection of cassava germplasm
- The cassava collection held in trust at CIAT includes a total of 6,155 accessions from 28 countries, represented in 5,690 clones of (*Mannihot esculenta*) and 465 genotypes of wild species conserved using *in vitro* techniques
- Over 37% of the cassava diversity held in CIAT's gene bank originates from Colombia, with another 24% coming from Brazil
- Collections from other South American countries (21%), as well as Central America and the Caribbean (7%), and Asia (7%) are also conserved at CIAT

3. Forages

- CIAT safeguards the greatest and most diverse tropical forages collections in the world, with 22,694 materials (127 genera and 700 species) to date
- This collection includes 21,083 legume materials and 1,611 grasses materials from 75 countries
- Between 1977 and 1993, 75 explorations were conducted that contributed more than half of the preserved material, increasing greatly the originality of the collection, while 9,877 materials from 41 countries were received as donations.

6. International Crop Research Institute for Semi-Arid Tropics (ICRISAT)

- The ICRISAT Gene bank serves as a world repository for the collection of germplasm of the six mandate crops: sorghum, pearl millet, chickpea, pigeon pea, groundnut, finger millet; and five small millets: foxtail millet, little millet, kodo millet, proso millet and barnyard millet
- With 128,155 germplasm accessions assembled from 144 countries through donations and collection missions, it is one of the largest international gene banks
- ICRISAT established regional gene banks at Nairobi, Bulawayo and Niamey to facilitate easy access of regional and global diversity to NARS partners in the region

- Together, the three regional gene banks conserve 43,353 accessions of mandate crops
- Several landraces now conserved in the ICRISAT gene banks have disappeared from their natural habitats in Africa and Asia
- The collection serves both as insurance against genetic erosion and a sources of tolerance to diseases and pests, environmental stresses, higher nutritional quality and traits related to yield for crop improvement
- The majority of collection have been placed in-trust with the FAO of the United Nations and the International Treaty on Plant Genetic Resources for Food and Agriculture (ITPGRFA) for use by the global community as International Public Good (IPG)
- ICRISAT Gene bank has deposited over 111,000 accessions at Svalbard Global Seed Vault (SGSV), Norway as safety backup
- ICRISAT gene bank has distributed over 792,000 samples of its mandate crops and small millets to users in 148 countries in addition to over 697,000 samples to scientists within the institute
- The collections held at the gene bank are also serving the purpose of restoration of germplasm to the source countries when national collections are lost due to natural calamities, civil strife, etc.
- ICRISAT gene bank has restored over 55,000 accessions to nine national programs in Asia and Africa
- ICRISAT gene bank has promoted testing and release of 104 accessions directly as 137 superior varieties in 51 countries and 1019 varieties were released in 81 countries utilizing germplasm and breeding lines from ICRISAT contributing to food security

7. International Food Policy Research Institute (IFPRI)

- The International Food Policy Research Institute (IFPRI), established in 1975, provides research-based policy solutions to sustainably reduce poverty and end hunger and malnutrition
- The Institute's regional and country programs play a critical role in responding to demand for food policy research and in delivering holistic support for country-led development
- For conservation of germplasm and exchange activities, IFPRI has developed genetic resource policies which focuses on understanding and developing policies to promote sustainable management of agricultural biodiversity by enhancing poor farmers' access to diverse crop genetic resources

- This program aims to better manage biodiversity and enhance agricultural productivity and rural livelihoods

8. International Institute for Tropical Agriculture (IITA)

- IITA has made a long-term commitment to conserve Africa's plant and animal genetic resources for future generations
- At IITA, expansive collections of both crops and non-crops, in (*in situ*) or out (*ex situ*) of their natural environment are conserved
- The collection along with plants, also include insect species which are included in Africa's rarest insect species category
- IITA's Gene bank holds over 28,000 accessions of plant material or germplasm, of major African food crops
- Started in the mid-seventies, the Gene bank helps in crop improvement and also provides "seeds of hope" for people affected by flood, fire, wars, and other disasters.
- The main crops stored in the Gene bank are **cowpea, cassava, plantain and banana, yam, soybean, Bambara groundnut, and maize**
- In addition, substantial collections of wild cowpea relatives and miscellaneous legumes have been collated over the past 30 years
- More recently a small collection of African yam bean, an underused legume, has been adopted for conservation
- However, the most important crop in the Gene bank is the **cowpea**
- The IITA Gene bank holds the world's largest and most diverse collection of cowpeas with 15,122 unique samples from 88 countries, representing 70% of African cultivars and nearly half of the global diversity. This incredible collection makes IITA integral in the protection of the cowpea species

9. International Livestock Research Institute (ILRI)

- The main objective of ILRI gene bank is to conserve, study and maximize uses of the biodiversity of forage genetic resources
- These activities are part of collaborative work financed through the CGIAR Gene bank platform
- The gene bank in Addis Ababa, Ethiopia conserves about 19,000 accessions from over 1,000 species
- This is one of the most diverse collections of forage grasses, legumes and fodder tree species held in any gene bank in the world. It also includes

the world's major collection of African grasses and tropical highland forages

- ILRI also manages field gene banks for grasses that rarely produce seeds or whose seeds are short-lived at Zwai and Debre Zeit in Ethiopia
- Seeds of ILRI's own collections are duplicated at the *Centro Internacional de Agricultura Tropical* (CIAT) in Colombia and at the Svalbard Global Seed Vault in Norway
- In 1994, the germplasm held by ILRI was placed in trust under the auspices of the Food and Agriculture Organization of the United Nations (FAO) as part of an international network of *ex situ* collections
- ILRI claims no ownership nor seeks any intellectual property rights over the germplasm and related information
- In October 2006, ILRI signed an agreement to include this material under the International Treaty on Plant Genetic Resources for Food and Agriculture
- Regarding the exchange activities, ILRI freely distributes about 3,000 samples of germplasm globally for evaluation and further development and use by smallholder farmers
- ILRI also maintains the Herbage Seed Unit which focuses on providing a source of tropical forage seeds and planting material of selected best-bet species at cost for use in establishing national forage seed production, including 33 species of herbaceous legumes, 10 species of grasses and 5 species of fodder trees
- The gene bank also provides group training for national program scientists in germplasm management and seed production and individual training for associates and interns

10. International Maize and Wheat Improvement Centre (CIMMYT)

- It manages most diverse maize and wheat collections
- CIMMYT's germplasm bank, also known as a seed bank, is at the center of CIMMYT's crop-breeding research
- This remarkable, living catalogue of genetic diversity comprises of over 28,000 unique seed collections of maize and 150,000 of wheat
- From its breeding programs, CIMMYT sends half a million seed packages to 800 partners in 100 countries each year
- CIMMYT's germplasm research aims to:
    - Conserve, characterize, distribute and use genetic resources

  - Safely distribute seed
  - Promote scientific stewardship and ensure open-access to CIMMYT's data and derived information
  - Create quality, open-source software
  - Develop and validate new tools and methods for gene mining and crop improvement
  - Build capacity in all the above areas
- Finally, the Seeds of Discovery (SeeD) project analyzes and documents maize and wheat biodiversity and facilitates its use in crop breeding to address current and future production challenges

11. International Potato Centre (CIP)

- The International Potato Centre (*Centro Internacional de la Papa)* is a research facility based in Lima, Peru, that seeks to reduce poverty and achieve food security on a sustained basis in developing countries through scientific research and related activities
- The crops which are the focus of this institute includes potato, sweet potato, other root and tuber crops
- It also focuses on improved management of natural resources in the Andes and other mountain areas
- It was established in 1971 by the Peruvian government

1. Potato

- Maintains 1,800 potato accessions comprising traditional cultivars
- Among the 45 years of existence of the potato gene bank, the number of cultivated potato accessions peaked at 17,326 accessions
- After extensive research involving the identification and removal of duplicates, the active cultivated potato collection now total 4,727 accessions including 4,354 traditional landrace cultivars from 17 countries (mainly from the Andean region) and improved varieties
- The entire clonal collection is conserved *in vitro* and distributed internationally as tissue cultured materials
- This Global collection is maintained in Trust and is distributed with the Standard Material Transfer Agreement (SMTA) under the terms of the International Treaty for Plant Genetic Resources for Food and Agriculture (ITPGRFA)

2. Sweet Potato

- The International Potato Center (CIP) maintains one of the world's largest cultivated sweet potato gene banks with over 5,500 accessions maintained *in vitro*
- The overall objective is to conserve the diversity in the collection and make it available to the global community for research, breeding, and training

12. International Rice Research Institute (IRRI)

- IRRI is an independent, non-profit, research and educational institute, founded in 1960 by the Ford and Rockefeller foundations with support from the Philippine government
- The institute, headquartered in Los Baños, Philippines, has offices in 17 rice-growing countries in Asia and Africa, and more than 1,000 staff.
- The International Rice Gene bank, maintained by IRRI, holds more than 130,000 accessions as of December 2018
- This includes cultivated species of rice, wild relatives and species from related genera
- The gene bank is the biggest collection of rice genetic diversity in the world
- In the past five years, they have distributed 159,815 samples to 1419 recipients in 61 countries
- The species of rice conserved in the International Rice Gene bank include:
    1. ***Oryza sativa*** or **Asian rice**
    2. ***Oryza glaberrima*** or **African rice** originated in West Africa.
    3. Twenty-four wild species of rice that are found in Asia, Africa, Australia, and the Americas.
    4. Nine species from seven related genera
- Each type of rice in the International Rice Gene bank is stored in both the base (-20 °C, long-term storage) and active (2-4 °C, for distribution) collections
- Viability testing of all samples is carried out every 5 years for seed stored in the active collection, or every 10 years for seed stored in the base collection
- If viability falls below 85%, a sample of the remaining seeds is planted to produce fresh seeds for storage.

13. International Water Management Institute (IWMI)

- The **International Water Management Institute** (IWMI) is a non-profit, scientific research organization focusing on the sustainable use of water and land resources in developing countries.
- IWMI is the lead centre for the CGIAR Research Program on Water, Land and Ecosystems (WLE)
- IWMI works in partnership with governments, civil society and the private sector to develop scalable agricultural water management solutions that have a real impact on poverty reduction, food security and ecosystem health.
- **IWMI's Mission- T**o provide evidence-based solutions to sustainably manage water and land resources for food security, people's livelihoods and the environment.
- **IWMI's Vision**- 'a water-secure world'. IWMI targets water and land management challenges faced by poor communities in the developing countries, and through this contributes towards the achievement of the Sustainable Development Goals (SDGs) of reducing poverty and hunger, and maintaining a sustainable environment
- Several tools have been developed in order to utilize the water resources efficiently

14. World Agroforestry (ICRAF)

- The ICRAF Genetic Resources Unit has a global role to collect, conserve, document, characterize and distribute a diverse collection of agroforestry trees, mainly focusing on indigenous species
- The ICRAF **seed bank** in Nairobi and **field gene banks** in the regions ensure the supply of superior tree germplasm for research and conserve material for the benefit of present and future generations
- The current aim of *ex situ* conservation activities at ICRAF is to be a world leader in the conservation of agroforestry tree germplasm and develop a global conservation system for priority agroforestry trees
- **Genetic resources databases** provide information on agroforestry tree taxonomy, uses, suitability and sources of seed as well as details of the ICRAF agroforestry genetic resources collection
- The Genetic Resources Unit of the World Agroforestry Center (ICRAF) conserves a wide diversity of about 190 species. These are wild, partially domesticated and domesticated trees used in agroforestry systems to supply fruit, timber, medicines and other products

- In addition, ICRAF maintains a unique collection of African dryland woody-fodder species
- Conservation of tree species poses special difficulties, including the large number of genera each with specific requirements such as long generation times and seeds that often cannot be dried for conventional storage
- As a result, ICRAF has to maintain field gene banks. These field gene banks both conserve diversity and offer a source of sexual and clonal reproductive material.
- Overall, 12,000 accessions of 67 diverse tree and shrub species are maintained in field gene banks spread across 42 sites in 16 countries
- About 19% of the seed accessions are safety duplicated at the Svalbard Global Seed Vault. About 8% of the accessions are maintained in long-term storage at Kunming Institute of Botany (KIB) in China.
- A database -- **the Agroforestry Species Switchboard** -- gives access to "information sources to support tree research and development activities"

15. World Fish

- WorldFish, an international, non-profit research organization, harnesses the potential of sustainable fisheries and aquaculture to increase food and nutrition security and promote better livelihoods
- Research is focused on the two interlinked challenges of sustainable aquaculture and sustaining small-scale fisheries, with integrated crosscutting themes of gender, youth, capacity development and climate change
- WorldFish focuses its expertise and research in the following areas:
  1. Building adaptive capacity to climate change in fisheries and aquaculture
  2. Strengthening gender equality in fish-dependent communities
  3. Increasing the benefits to poor people from fisheries and aquaculture value chains
  4. Improving nutrition and health through fisheries and aquaculture
  5. Identifying and promoting policies and practices to increase the resilience of small-scale fisheries
  6. Sustainably increasing the productivity of small-scale aquaculture

References

Halewood, M., Jamora, N., Noriega, I. L., Anglin, N. L., Wenzl, P., Payne, T. and Lusty, C. (2020). Germplasm acquisition and distribution by CGIAR genebanks. Plants, 9(10), 1296.

Halewood, M., Gaiji, S. and Upadhyaya, H. D. (2004). Germplasm flows in and out of Kenya and Uganda through the CGIAR: A case study of pattern of exchange and use to consider in developing national policies.

Kumar, P. L., Cuervo, M., Kreuze, J. F., Muller, G., Kulkarni, G., Kumari, S. G. and Negawo, A. T. (2021). Phytosanitary interventions for safe global germplasm exchange and the prevention of transboundary pest spread: the role of CGIAR germplasm health units. Plants, 10(2), 328.

Robinson, J. (2010). Scoping Study: The Impact of Germplasm Collection, Conservation, Characterisation and Evaluation (GCCCE) in the CGIAR.

Robinson, J. and Srinivasan, C. S. (2013). Case-studies on the impact of germplasm collection.